EXPLORING GENOMES

WEB-BASED BIOINFORMATICS TUTORIALS

www.whfreeman.com/young

PAUL G. YOUNG
QUEEN'S UNIVERSITY

W. H. FREEMAN AND COMPANY
NEW YORK

Cover Design: Rae Grant
Cover Image: © Alfred Pasieka/Science Photo Library/Photo Researchers

ISBN: 0-7167-5738-9

Printed in the United States of America

Second Printing, 2003

TABLE OF CONTENTS

Introduction

Computer science has had a major impact on the biological sciences, and this is particularly true in the field of sequence analysis. As sequencing technology improved and became highly automated in the 1990s, researchers around the world accumulated a wealth of information that rapidly grew beyond the scope of what scientists could analyze independently. Scientists and government officials with considerable foresight lobbied for centralized institutions that could store this information and make these resources available to researchers worldwide through the internet. At the same time, programmers were developing potent analysis tools that could mine the information in the databases for comparisons at a very detailed level. With the advent of easy and near universal internet web access, researchers have come to rely on these institutions to an ever increasing extent. *Genomics*, the study of whole genomes, and *bioinformatics*, the development and use of computer tools to analyze them, now contribute to virtually all areas of biological science. It is safe to say that understanding how to make use of these resources is an essential skill for anyone in the fields of biology and biomedical sciences.

The resources available are located on thousands of different web sites and are quite varied in size and nature. Some are very narrowly focused, such as databases created for a particular organism or sequencing project. Alternatively, a site might be set up simply to allow the use of a particular type of analysis or a new software tool. Some sites may represent the efforts of a single laboratory. Others, such as the National Center for Biotechnology Information (NCBI) in Washington, D.C., have as their mandate the collection of all publicly available sequence and related information and the development of software tools for its use.

The curators and computer programmers at comprehensive centers such as NCBI compile sequence and related data as it becomes available from sequencing centers around the world and present it in a form that is easily accessible to researchers. Their efforts are coordinated with similar genome centers in many countries. NCBI and its counterparts, such as the European Molecular Biology Laboratory (EMBL) in Germany, and the DNA Database of Japan (DDBJ), are the primary centers for the collection and archiving of sequence data.

The software tools that are used for access and analysis differ at the various web sites. First and foremost, an archival site such as NCBI allows researchers to retrieve the original sequence record and all subsequent updates and modifications for a gene, protein, or genome. The sites also provide tools for gene alignment that are essential in answering questions regarding the conservation of genes among organisms. This often allows information from major genetic systems such as yeast, *Drosophila,* or *Arabidopsis* to provide clues as to the biological function of a human gene, or vice versa. At another level, the programmers actively participate in designing new software tools for problems as diverse as identifying the coding regions within newly sequenced genomes, for linking gene discoveries to human diseases, and for examining and comparing the three-dimensional structures of proteins.

Students trying to find their way through one of these web sites are often intimidated by the unfamiliarity of the material and terms, as well as a sense that their computer skills are not as advanced as they would like. They needn't worry on either count. Web-based access at its best is designed to allow very sophisticated analyses to be carried out with little or no detailed knowledge of the underlying programming. All the student needs is a little guidance.

The tutorials in this book and on its accompanying web site at **www.whfreeman.com/young** focus almost entirely on a single site, the National Center for Biotechnology Information. It is one of the world's largest resources and it has a wide range of useful tools and presentations available. The tutorials here are designed to introduce the basics of genomic analysis at a level appropriate for college students. These tutorials give students initial guidance and practice in exploring the breadth and depth of the resources by helping them understand the structure and content of the databases, and facilitating direct practice in the use of search and alignment tools.

Each tutorial is written as an interactive guide to an aspect of the NCBI site. The user's browser window is split into two parts. Guidance is provided as text on one side of the screen, while the main window is directly connected to NCBI. This latter window runs live on the NCBI computers. Using real examples from the research literature, the tutorial walks students through the exercise and tells them precisely how to run the programs and interpret NCBI results.

These tutorials are not intended to be comprehensive. In many ways, they simply scratch the surface of the resources. They do, however, allow a student to gain practical skills and to do independent research very quickly. A student that becomes familiar with this material will have no problem going forward to more advanced use of the resources. NCBI itself has a broad array of tutorials associated with their web site. Many of these are at a more advanced level and the introductory ones provided here serve as a bridge to them.

The tutorials are designed to display some of the major public resources of the NCBI, keeping in mind the level and interests of the beginning genetics student. For each tutorial the searches are run with known genes or sequences and for the most part examples from our major genetic systems such as *E. coli, Drosophila*, yeast, *Arabidopsis,* and human. Usually a fairly common protein or gene is chosen so that it might be familiar to a student in genetics. The choice of example is sometimes dictated by a desire for a relatively simple result (not always possible).

The tutorials include an overview of the NCBI site as well as examples of how to use the information resources such as Entrez, PubMed, OMIM, and the Cancer Genome Anatomy Project. There are several tutorials dealing with gene alignments using BLAST, as well as searches for conserved domains using pattern and profile searching. The student can explore the 3D structure of nucleic acids and proteins and even search the database for proteins with similar 3D structure. Lastly, taxonomy resources are examined and a multiple alignment of a gene family is used (ClustalW) in order to draw a phylogenetic tree. In all cases, students are asked to go back to the program using an input of their choice (with some suggestions made) so that they can reinforce their knowledge.

Database resources such as NCBI and the European Molecular Biology Laboratory (EMBL) are dynamic and constantly being improved. They increase in size daily and frequently add to the services that they offer. The inevitable consequence is that occasionally hyperlinks are altered. Although these tutorials are frequently monitored to ensure that they work, sometimes a glitch occurs. We would appreciate hearing of any problems so that they can be corrected. Please email us at **techsupport@bfwpub.com**.

Detailed Table of Contents

other resources) and allows us to examine the various entries organized by organism and their position in the phylogenetic tree. Sequence data allows us to quantitatively estimate the evolutionary distance among organisms based on the extent of sequence divergence. We will examine this using a gene that is highly conserved in all organisms. By doing a multiple alignment of the sequences using the Clustal and Phylip programs, we will be able to estimate the evolutionary distance among the various organisms and to draw a simple phylogenetic tree.

1. Introduction to Genomic Databases

Starting link: **http://www.ncbi.nlm.nih.gov.**

Everyone has a difficult time keeping up with the flow of new information. This is particularly true in biology now as the pace of discovery accelerates. Databases have become an essential tool for accumulating and archiving raw data. They also play a major role in analyzing and presenting information to researchers and the public in an easily accessible form. In this *Exploring Genomes* tutorial we will survey one of these resources: The National Center for Biotechnology Information (NCBI) located in Washington, D.C.

One of the roles of NCBI is to archive raw DNA sequence data. The sequence information comes from research efforts in laboratories around the world as well as from large-scale, dedicated genome sequencing centers. The resulting database is referred to as GenBank. GenBank shares its resources with the European and Japanese equivalents so that there are three primary public repositories of such information in the world. Because of the automation of DNA sequencing over the past decade, these databases are increasing in size exponentially. GenBank includes the sequences of the *E. coli*, *Drosophila*, and human genome, as well as data from thousands of other species. At present it comprises some 10 million DNA sequences with a cumulative length of 11 billion base pairs. You can see the rate of growth by going to **http://www.ncbi.nlm.nih.gov/Genbank/genbankstats.html.**

Now let's go back to the NCBI home page, the public access point to its many resources (**http://www.ncbi.nlm.nih.gov/**). A quick glance at this page shows that NCBI contains far more than just the sequence repository. It is a rich source of information on all aspects of genetics and genomics. All of the divisions are searchable and information ranging from gene sequences, to the position of a locus on a human chromosome, to direct access to the scientific literature dealing with a particular gene, is immediately accessible.

There are six major categories of service listed across the top of the NCBI homepage. These are PubMed, Entrez, BLAST, OMIM, Taxonomy, and Structure. In addition there are specialized projects and databases listed on the right hand side of the page. In this *Exploring Genomes* tutorial we will take a quick look at some of these resources. Subsequent tutorials will explore a particular resource in much greater depth. Let's click on each of the six resources at the top in turn. Click on **'PubMed'** first.

PubMed is the NCBI gateway to the biomedical research literature. It is a searchable database and information can be retrieved based on combinations of parameters such as author, subject key words, or organism. A complex query can be entered and a list of publications matching it will be returned. For instance, we could enter a simple search by author. If we entered **Hartwell LH** (one of the winners of the 2001 Nobel Prize in Medicine) and pressed **'Go'**, PubMed would return a list of his current publications. Give it a try.

The list of Hartwell's publications is 5 pages long. Only the top and therefore most recent papers are displayed on the first page. Each paper is linked to its abstract and sometimes the full text of the articles. We will explore PubMed much more fully in a later tutorial. Let's go back and try another NCBI division by clicking the **NCBI icon** on the top left to get to the NCBI homepage again, and then clicking the button for **'Entrez'**.

The Entrez button opens a search engine that links all of the NCBI databases together. This allows access to everything from sequence to structure. The options are in the drop-down window at the left. The PubMed link that we looked at initially was only one of these options. Let's try searching for the human keratin protein sequence in the Protein database. Choose **'Protein'** from the drop-down menu, and then type **'keratin AND human'** in the text box. Now press **'Go'**.

The search returns a list of database entries including keratin-associated proteins and keratins themselves. Note that only the first 20 entries of around 500 are displayed on the first page! For each entry, clicking on the associated links will display the sequence, related information, and links to other parts of the databases. We can explore these later. Let's go back to the **NCBI** homepage and try the **'BLAST'** button.

BLAST is a powerful nucleic acid or protein alignment tool. It allows us to dynamically search the sequence databases (all 11 billion base pairs!) to find similar sequences in different organisms. It is extremely versatile and comes in many different forms for doing different types of searches. The underlying method is the same in each case, however. This is our standard software tool for doing such searches. It is very important and we will devote an entire tutorial to its use later. For now, let's move back to the **NCBI** homepage and click on to **'OMIM'**.

OMIM is the Online Mendelian Inheritance in Man database. Note from the overview on the OMIM homepage below that it integrates the known Mendelian genetics of human disease with the resources made available through Entrez at NCBI. We will explore this in detail later but for now let's open the **OMIM Statistics** link, under OMIM Facts on the left-hand menu.

The OMIM statistics page gives an overview of the breadth of the resource. Note the categorization by Mendelian inheritance pattern. Note also that over 13,000 entries are available with at least some information. Keep in mind that we believe there are more than 35,000 human genes! All genes will not necessarily be associated with a disease or visible phenotype and therefore for many genes little information is yet available. The Human Genome Project has provided a glimpse at large numbers of new genes of unknown function, reminding us of how much remains to be done.

Next, click back to the **NCBI** homepage and on to the **TaxBrowser** database.

The Taxonomy section groups all data by taxonomic classification. You may type in a species name to find out if any sequence information is available; for the major genetic systems, simply click on the links on the Taxonomy Browser home page. Try *Caenorhabditis elegans*, a nematode worm, one of our most powerful model genetic systems whose full genome sequence was recently determined.

This organism-specific page gives important information regarding phylogenetic lineage as well as the number of sequences of various types that have been deposited. These groupings may be called up at will and each of the genes listed are linked to the Entrez system. One important use of the groupings is to restrict other types of searches. For instance a BLAST search can be launched from the BLAST page and the database searched restricted to *Caenorhabditis elegans* only (or any other species).

Now, let's go to the last major subdivision, '**Structure**'.

The structure database contains the 3D structure for all nucleic acids and proteins whose shape has been determined by X-ray crystallography or nuclear magnetic resonance. We can call up these 3D models at will. The structure database is associated with the VAST program that allows for 3D structural comparisons among different proteins. It also will search the database on the basis of structure. These are very powerful tools and we will give them a try later in the term. Now, let's go back to **NCBI** home page to see some of its other resources.

Apart from the basic databases and access software, NCBI has a wide variety of highly specialized databases and analyses. These focus on particular problems or interest groups. Some are listed on the right hand side of the NCBI homepage below. We will just touch on a couple of them for now to get a sense of their capabilities.

The first is the Human map viewer. This allows us to visualize the various human chromosomes and the genetic loci on them. Let's take a look by clicking **'Map Viewer'** under the Hot Spots list. From the resulting page, choose **'Homo sapiens (human)'** from the Mammals category.

The human genome view is a visualization of the full chromosome set. Clicking on any chromosome number will expand the view of that particular chromosome. Let's try the **Y** chromosome.

The graphic of the Y chromosome is expanded so that we can see its banding pattern as seen by a cytologist. Aligned along it are the blocks of DNA that have been sequenced and represented here (contigs) followed by the genes located on these pieces of DNA. The genes are represented as blue diamonds in the second column labeled "Uni...", which stands for Unigene, the NCBI database that contains the compiled summary data for each gene. Roll over each diamond to get the identification information for each gene. Subsequent columns provide links to the various genes that have been annotated on the DNA sequence. Click on any of the **blue diamonds** in the Unigene list.

The UniGene summary is an annotated view. It is compiled by a curator and takes into account all known information regarding the sequence. UniGene also includes links to OMIM as well as a comparison to the most similar genes in other organisms. It is a very rich source of information. All of the data is cross-referenced by links into Entrez and various other databases that provide the gene sequence and research literature references. Ultimately the entire human genome will be available in finished form from this database.

Now let's go back to the **NCBI** homepage and click on the first resource under Hot Spots, **'Cancer genome anatomy project'**.

The Cancer Genome Anatomy Project is the last topic we'll briefly explore. This database focuses only on tumor tissue and strives to provide cross-referenced information for all genes thought to be involved in cancer. Each of the subsections enters the database with a different type of approach. Ultimately you can explore the database to great depth, searching out the genes involved, chromosomal locations and aberrations, and biochemical pathways. All of this information is related to the tissues and cells where the tumor that you are interested in originates.

Over the course of these *Exploring Genomes* tutorials, we will look at parts of these databases in much greater detail. A facility for handling them is an essential skill for a modern biologist. Ultimately, the only way to familiarize yourself with a resource of this type is to go to the web site (**http://www.ncbi.nlm.nih.gov/**) and start exploring some of the links. You might start by searching for the answer to the following simple questions:

- What is the publication history of your Biology and in particular your Genetics instructors?
- What organisms do they work with?
- What types of question are they trying to answer?

2. Learning to Use Entrez

Starting Link: **http://www.ncbi.nlm.nih.gov**.

The Entrez retrieval system is the entry point for searching most of the material at NCBI. Its strength is that it provides links between related types of information. For instance, in storing a DNA sequence file, the file is associated with the protein translation of the sequence, with the literature reference, and with links to similar genes or proteins in other organisms. It also provides a characterization of any notable features, such as conserved regions, and ultimately chromosomal location and 3D structure of the gene product. The retrieval system moves relatively effortlessly among these various types of data. The links are updated as new data are added to the databases (or new databases are developed) and the resource becomes richer and richer over time.

We looked very briefly at Entrez in the introductory tutorial. Now let's learn to use it by walking through an analysis of the dystrophin gene in humans, which is responsible for causing Duchenne muscular dystrophy (DMD).

Click on **'Entrez'** on the top blue bar to access the Entrez home page from the NCBI home page.

Remember that in the drop-down window on the left (in the 'Search' box) there are several options indicating different databases. All of these databases are ultimately linked within the framework of the Entrez retrieval system.

Let's start with a research literature search for dystrophin on PubMed. Since it is a well-characterized protein involved in an important human disease, expect a long list of retrievals.

Choose **'PubMed'** from the Search drop-down menu, type **'dystrophin'** into the text box, and then click **'Go'**.

The search returns a rather daunting several thousand entries, obviously a very active research area. They list in reverse chronological order and only the last few are displayed here. Notice the titles of some of the papers. Clicking on the authors' names will link you to the publication information and abstracts. Note that the reference numbers change as new publications are added to the database.

For this exercise, we are going to take a look at a paper written in 1987, when the molecular genetics of the dystrophin locus and the Mendelian genetics of Duchenne muscular dystrophy were converging.

Let's call up the Hoffman, Brown and Kunkel (1987) reference. Since it is so old, and the articles are listed in reverse chronological order, it will take a lot of clicking through the pages to get to it. A faster way to reach the article is to do a search with limits. Click **'Limits'** on the light blue bar above the Display button. In the Publication Date field From, type in **'1987 01 01'**. In the field To, type in **'1987 12 31'**. Now press **'Go'**.

When the search is completed, you should see the paper by Hoffman, Brown and Kunkel (1987) near the top of your screen. Now click on the **authors' names** to link to the information related to this paper.

The abstract of the paper is displayed as well as a link to the full text article. For most articles there is a registration and charge to go beyond this point. Also included is a set of links on the right (in blue). These point to closely related research articles, to the gene and inheritance pattern, as well as books and other resources. Clearly we could pursue an extensive library research project related to this gene simply by following a number of these links. For the moment we will find the chromosome map of the gene responsible for DMD.

From this page, click on **'OMIM'**, one of the light blue links near the title of the paper.

This link will connect us to all the literature related to DMD and dystrophin. For now, though, we just want to see the gene map locus. Click on **'Xp21.2, 12q21'**.

Of the map locations that appear next, click on **'DMD, BMD Dystrophin (muscular dystrophy, Duchenne and Becker types)'**.

Now you should see a chart with all the gene locations related to DMD and dystrophin. Click on the first location, **'Xp21.2'**. This will take us to Entrez Genome page.

The graphic that appears is a view of the human X-chromosome. The DMD (Duchenne Muscular Dystrophy) gene is highlighted. Clicking on the DMD link itself will take us to the LocusLink page where there is an extensive compilation of information about the gene and its expression. Click on **'DMD'**.

LocusLink is a very rich resource of data (scroll down through the pages and examine it) that links back to the other databases at NCBI. By now you should be getting a sense of the extent of the interconnection built into the Entrez structure. Clearly we could pursue this gene in many directions from this page. However, let's go back to the Entrez home page and initiate our search for dystrophin by another route.

Click **'Entrez'** on the black menu bar. From the Entrez page we could start our search a different way with a search for the Protein record for dystrophin. Choose **'Protein'** from the Search drop-down menu, type **'dystrophin'** into the text box, and then press **'Go'**.

The output is very long with hundreds of entries, including various alleles of dystrophin and many dystrophin-related proteins. The sequence we are interested in has the identifier P11532. This identifier number is the **accession number** and was permanently attached to the sequence file when it was first archived in the database.

You can find P11532 by clicking through the pages of results until you see it, or you can do another search in the Protein database for **'P11532'**.

When you find the sequence, you will see that on its right are a number of links. These include BLink, which compares the sequence to others in the database, a link to conserved domains in the proteins, and a drop-down menu for several other links. Keep in mind that many of these links are displaying different aspects of the NCBI information in different ways. The underlying databases are the same. Let's click on **'BLink'**.

The BLink program displays a graphic of an alignment of our gene with all related sequences in the database. Note that there are many dystrophin entries and the graphic on the left shows the regions of similarity in sequence. Some of these are different isoforms and alleles; some are from other species. Some are older records. NCBI is an archival database, and once generated, all records are kept, although they can be updated. All of these records are available from this page simply by clicking the links. A researcher would use this resource to examine the various records for the gene, and to see, in a simple graphical way, different forms of the protein in humans, or the extent to which dystrophin from other species is similar to the human form.

From here, let's look at the dystrophin protein record itself by clicking on the link **'A27605'**, the first entry under the Accession number column.

The page that comes up is the protein record for our accession number A27605. This record is a computer translation of the DNA sequence determined by sequencing the mRNA for the protein. Computers require standard formats to store data, and humans need accessible formats to view it. This file (in a human readable form) is a **GenBank flatfile.** The display you see is the standard output format for GenBank. It has a specific list of sections, each with a particular type of information. It also includes highlighted links to other aspects of the databases. We don't have to worry about them all, but let's look at a few.

The locus is identified by the accession number, at the top of the record. The DEFINITION line shows it as "dystrophin, muscle – human".

Scroll down to see the REFERENCE section, which gives us all of the literature references related to the sequencing and characterization of this protein.

Further down, the FEATURES list tells us about various predicted or biochemically verified conserved domains, regulatory sites, and binding sites for other proteins. The features are arranged in order starting at the amino terminus of the protein (designated 1) and their location is identified by the amino acid position along this very large protein (3685 amino acids).

Lastly, at the bottom of the record is the protein sequence itself in single letter code.

The format of the GenBank flatfile is a rich source of information and you should familiarize yourself with it. There are other file formats, however, that are much simpler and sometimes more appropriate to use, particularly if all you want is the sequence for insertion into another program. Some analytical programs will accept the accession number as input, but many require that you cut and paste in the sequence. One of the simplest and most widely used formats is the **FASTA** format, which was developed for one of our important search and alignment programs (FASTA).

We can see the FASTA version of our dystrophin protein by toggling the display drop-down menu (at the top left side of the page) to **'FASTA'** and pushing the **'Display'** button.

The format has one line of information beginning with the > sign and followed by the accession number and species information. Starting on the next line is the sequence of the dystrophin protein. Notice that there are no line numbers or punctuation – it is perfect for cutting and pasting.

Now let's go back and learn how to execute more specific searches in Entrez. Close the window that we've been working in, and then go back to the Entrez home page at **http://www.ncbi.nlm.nih.gov/Entrez/**.

Oftentimes in searching for information on a protein such as dystrophin, we can be overwhelmed with the hundreds or thousands of responses. If you have a more narrow interest it is possible to phrase searches using Boolean operators (AND, OR, NOT). For example if we searched the protein database for "dystrophin AND chicken" then our output is sharply focused. We only get a small number of entries including dystrophin and some dystrophin-related proteins from chickens.

See how many results you get when you select the **Protein** database, write **'dystrophin AND chicken'** in the text box, and then press **'Go'**. (Note: The Boolean operators must be written in ALL CAPS.)

There are other ways to limit searches as well. For example, since there are so many human records for such an important human genetic disease gene locus, we might want to exclude human data from the list. Search the protein database for **"dystrophin NOT human"** to simplify the output. Now we get everything but human and the list shrinks from thousands to a few hundred entries.

There is really only one way to become conversant with such an important resource. Go to the NCBI Web site at **http://www.ncbi.nlm.nih.gov/** and continue to explore the Entrez program. This is one of the most useful tools for any biologist and could be considered an essential skill.

You have already examined the publications produced by your professor. Choose one of those publications related to a particular gene, protein, or organism and search through Entrez to find the associated links. Start your search from the publication you've chosen in PubMed. Using the links on the records, go to Entrez and input a gene or protein name that you have gleaned from one of the abstracts of the papers. This will initiate your search – see how far you can go!

3. Learning to Use BLAST

Starting link: **http://www.ncbi.nlm.nih.gov**.

This tutorial is a more detailed look at the BLAST (Basic Local Alignment Search Tool) program at NCBI, the most important single software tool for searching sequence databases. BLAST can be used to search databases using nucleic acid or protein query sequences.

We will walk through two protein examples in this tutorial. The first is the protein insulin, which should be familiar to you for its role in the regulation of blood sugar and in diabetes. The second is dystrophin, which we examined in the Entrez tutorial, a large and complicated protein.

To get to the BLAST homepage from NCBI, click on '**BLAST**' at the top menu. This page is the starting point for several BLAST programs available from the BLAST homepage. Since insulin is a protein sequence, we will search using BLASTP. Click on '**Standard protein-protein BLAST [blastp]**' under the Protein BLAST heading.

The BLASTp page that appears contains a window for pasting in the query sequence as well as several optional parameters that can be set.

For our first database search we are going to use the insulin protein from the Zebra fish (*D. rerio*). The sequence of the insulin precursor protein in FASTA format is written as:

>gi|12053668|emb|CAC20109.1|insulin[Danio rerio]
MAVWIQAGALLVLLVVSSVSTNPGTPQHLCGSHLVDALYLVC
GPTGFFYNPKRDVEPLLGFLPPKSAQETEVADFAFKDHAELIR
KRGIVEQCCHKPCSIFELQNYCN

You should copy and paste the sequence above into the **Search** text box. Alternatively, you could do this search by typing the accession number into the window. Blast will recognize it and retrieve the sequence from the database.

The FASTA format has an identification line followed by the primary amino acid sequence in single letter code. The mature insulin molecule is a processed version of this primary sequence. It is derived by limited proteolysis from the initial translation product shown here.

The remainder of the search form allows you to set various options.

Set subsequence allows you to search with a particular portion of the sequence. Leave it blank so that the entire sequence will be used in the search.

Choose database has a drop-down menu that allows you to choose which part of the database you want to search – the entire database (nr for non-redundant), or sections of it such as *Drosophila* proteins only. Leave it on 'nr', the complete protein database.

Do CD-Search allows a comparison of the query sequence to a database of conserved domain patterns. This is a powerful tool for finding functional domains in genes. Leave it toggled on.

Last, click the **BLAST!** button to start the search.

Pushing the BLAST! button sends the file to the NCBI computer for the search . The page that appears next tells you that the sequence has been submitted and gives you a Search ID number.

At the top is the result of the **Conserved Domains** search. Conserved domains are the functional modules of proteins. They might include a pattern of amino acids typical of a particular catalytic site, or perhaps the binding site for a regulator of a protein.

Click on **Insulin** to see the full list of conserved domains. These will appear in a pop up window. Our search identifies two known domain patterns: insulin and insulin-like growth factor. The colored bars can be activated by rolling your mouse over them, so that the identification of the pattern shows up in the window. **Roll over** the bar for insulin. You should see the text in the box above the bars change to give the identification number of the pattern match, the fact that it is for insulin, an alignment score (S) indicating how strong the match was (higher is better), and a statistical measure of the significance of the match (E). The E value is the expectation that the match would have been found in the database by chance alone (lower is better). In this case, E is very small indicating that our query sequence is likely insulin (but we already knew that!). **Close** the pop up window.

Now, click the **FORMAT!** button to tell the computer at NCBI to send the results of the actual search.

The output of such a search is many pages long and is made up of several components.

The first major section is a graphical display of the strongest matches to the query sequence. Notice that some hits are to the full-length insulin precursor and some are to the shorter processed form. They are color-coded according to the alignment score. If you roll your **mouse over** the various lines, the identification information, alignment score (S), and E value appear in the text box above the chart.

The second section of the results is a detailed list of hits ordered by their alignment scores. They correspond to the ones displayed graphically. Note that each line gives the identification information for the protein followed by the alignment score and the E value. The entries are ranked from the lowest to the highest E value, which can be interpreted as from most similar to more distant. The top line, not surprisingly, is the record for Zebra fish insulin itself. If you click on the gene identifier link, it will call up the sequence from Entrez. When you click on the **Score** link, it will show you the particular alignment from lower down in the output file. You can return to this page by using the **back** button on your browser.

Further down the list, the output gives the actual gene alignments for the various proteins in the list we just looked at. The first of these represents the top alignment. It is a perfect match to the subject, the Zebra fish insulin record. Note however, that early in the sequence is a masked region (shown by XXXX). This is a low complexity, or repetitive, region that is masked out in the query and ignored in the database search. Such regions may interfere with the alignment. You can see the actual masked sequence, since it is the same protein, in the subject line (LLVLLVVSSVS); it is a mostly hydrophobic, repetitive sequence.

Now let's look at two examples with a less than perfect match.

Scroll down the list of sequences producing significant alignments until you find the *Xenopus* sequence (gi|124513|sp|P12706|INS1_XENLA), and click on its score, '**90**'.

The alignment with *Xenopus* insulin (South African clawed toad) is a less than perfect match. Many regions are still strongly conserved however (notice the low E value). Empty spaces indicate mismatches and a + sign indicates similarity between the two different amino acids compared. For instance, at the beginning of the sequence, a hydrophobic leucine is scored as similar to a hydrophobic valine. Notice also that a gap had to be inserted in the *Xenopus* sequence to give a good alignment with the query. This would be the site of an insertion or deletion event during evolution.

The last example that we will look at is the alignment with human insulin. **Scroll** back up to the list of sequences producing significant alignments, and look for gi|4557671|ref|NP_000198.1|, and click on its score '**79**'.

Notice that there are many records for human insulin in GenBank. The alignment itself is still very strong (note the E value). But, compared to the *Xenopus* alignment, there are several more gaps placed in both the query and subject, not surprising for more distantly-related sequences.

We are finished looking at this sequence, and we're going to search for another sequence. So, **close** the window that the Zebra fish results appeared in, and you should just have the tutorial window open.

For our second BLAST search, we will use the dystrophin protein that we examined in the Entrez tutorial. Dystrophin is a very large protein (several thousand amino acids) with multiple conserved domains. It is also very highly conserved within the vertebrate group.

To run the search let's go back to the BLASTP page by clicking on the '**Protein**' link at the top of this part of the page.

Now, we could call up our dystrophin accession number (P11532), choose the FASTA output and cut-and-paste the sequence into the search window. It is much simpler, however, to simply type in the accession number since in this case we know it and the NCBI BLAST program will accept it.

Try entering '**P11532**' in the Search box and pressing '**BLAST!**'.

We see the output from the Conserved Domains database. . This is a database of evolutionarily conserved patterns of amino acids. These small patterns identify the amino acid residues that are almost always conserved through evolution in particular functional domains. There are two types of conserved domains in dystrophin as shown in the graphical representation. The domain at the far left is the calponin homology domain.

Open the list of conserved domain results by clicking on it. Calponin is an actin binding protein in the cell. Note the E value in the window by rolling over the blue blocks that say 'CH'. The red and blue tags simply denote hits in two different Conserved Domain databases. Each database uses a somewhat different pattern., hence the different E values. The remainder of the domains marked are spectrin repeat domains.

Spectrin is an important protein of the cell's cytoskeleton. Dystrophin has 6 domains showing similarity to a motif in Spectrin. These are smaller and/or have weaker similarity than the calponin domain. Compare the E values yourself. The various domains are allowing you to see the modular nature of proteins.

Now, let's go **back** to the original results page and click **'Format!'**.

Once we receive the search result, it is obvious that there are many similar proteins in the database. Moving the **mouse over** the graphic will display the information line for dystrophin for a variety of vertebrates, a large number of different human dystrophin alleles, and sometimes gene fragments or other proteins aligning with specific domains. Now let's look at the subject hit list below the chart.

Note the E values for many of these subject hits on the right. The expectation is not exactly 0, of course, but rather is very close to a 0 chance of it being by chance alone, and has been rounded off. Note that the first real score is a rather small number, so you can imagine how small the other expectations must be! If we click on one of the scores, we can see the alignment. Let's try the mouse gene by clicking the score of '**1513**' for 'gi|192972|gb|AAA37530.1| (M18025) dystrophin [Mus musculus]'.

The mouse sequence is obviously very similar to that of the human but there are lots of positions where they have diverged. Given longer periods of evolutionary separation, we would expect to see more divergence.

We are finished looking at this sequence, and now we're going to try a more targeted search for another sequence. So, **close** the window that the dystrophin results appeared in, and you should just have the tutorial window open.

Let's now ask whether *Drosophila* has a dystrophin sequence. One way of focusing the search is to specify a smaller database. In this case we can search the *Drosophila* genome alone.

Go back to the BLASTP home page by clicking on **'Protein'** at the top of page. Paste the dystrophin accession number, **P11532**, into the Search text box. Then, in the 'Choose database' drop-down menu, select **'*Drosophila* genome'**. Alternatively, under the Options for advanced BLAST, choose **Drosophila melanogaster** from the 'select from' drop-down menu. Now click **'BLAST!'**.

When the results are in, click the **'Format!'** button to get the hit report.

Here the result is not as complicated as for the entire GenBank database. Clearly there are some very good scores, as well as various shorter alignments. Perhaps some of the latter represent calponin or spectrin repeat domains. Let's have a look at the top score by clicking on the **first number** underneath the 'Score (bits)' column.

The *Drosophila* dystrophin gene product is obviously quite strongly diverged from the human version. Here our overall score is very good but the sequence comparison shows only about 40% identities. Nevertheless we would expect, if we examined the conserved regions carefully, that they would represent the important functional domains of the protein. At this level of analysis, we are perhaps not so different from the fruit fly.

You have been investigating the publication record of your genetics professors and searching for information on a gene of interest to them. Go to the NCBI homepage (**http://www.ncbi.nlm.nih.gov/**), find the BLASTP program and do a BLAST search on one of the genes that you found.

4. Using BLAST to Compare Nucleic Acid Sequences

Starting link: **http://www.ncbi.nlm.nih.gov**.

In the previous tutorial, "Learning to use BLAST", you learned how to use BLAST to search the genomic databases and to compare protein sequences. Proteins have a high complexity and information content since there are 20 different possible amino acids at each position in the sequence. As we have seen, there are also groups of amino acids with related properties that are scored as being similar. Nucleic acids, on the other hand, have only four possible choices at each position (AGCT[U]). For the most part, we only look to match identical residues at a particular position in a DNA or RNA molecule. In this tutorial we will start using BLASTN to compare the sequences of different transfer RNA molecules.

From the NCBI home page, choose '**BLAST**' from the top menu, and then choose '**Standard 'nucleotide-nucleotide BLAST [blastn]**'' from the list underneath the heading "Nucleotide BLAST".

Transfer RNA molecules have a complex tertiary structure that is essential for their function and is dependent on intramolecular complementary base pairing. The requirement that tRNAs maintain this structure in order to interact with the ribosome and with aminoacyl tRNA synthetases results in strong selection in favor of retaining the primary sequences through evolution.

The following figure shows the cloverleaf stem-loop secondary structure of a tRNA molecule. This conserved structure is common to all tRNA molecules. In the diagram, we can more clearly see the complementary base pairing in the stems. Ribosomal RNAs that serve a structural role in the ribosome also show strong conservation of some regions. As we will see later, however, the constraints on the structure of messenger RNA sequences are less and there is correspondingly less conservation of sequence.

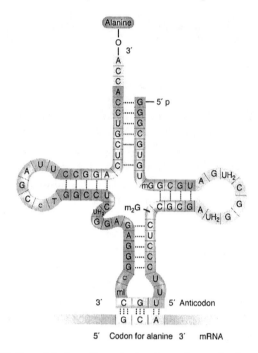

Figure 3-21 from *Modern Genetic Analysis, Second Edition* by Anthony J. F. Griffiths et al, © 2002 W.H. Freeman and Company.

BLASTN is very similar to BLASTP but obviously will only accept nucleic acids as input sequence. Cut and paste the phenylalanyl tRNA from *Drosophila melanogaster* into the **'Search'** text box.

>gi|174308|gb|K00349.1|DROTRF2 D.melanogaster phe-trna-2
GCCGAAATAGCTCAGTTGGGAGAGCGTTAGACTGAAGATCTA
AAGGTCCCCGGTTCAATCCCGGGTTTCGGCACCA

Just as for BLASTP, the program is run by pressing the **'BLAST!'** button and then the **'Format!'** button when it appears.

A search against the entire database at GenBank yields many hits, since tRNA molecules are highly conserved. The output list from this search shows many strong hits with low E values. Not surprisingly, many are from *Drosophila*—but also note other species, including human, in the list.

Look at the alignment for one of the human subject sequences by clicking on the Score '137'. You should see that it is almost identical to the query sequence.

Now let's look at the source for many of these sequences by clicking on the GI or accession number of any of the sequences. You will find that they are mostly in very large genome sequence files containing multiple genes; the tRNA subject sequence is somewhere internally in the DNA fragment.

Let's look at more distantly related hits, which would be much further down the list on the previous page. To illustrate a hit with low identity, first you should **close** the window that opened with the first search results in it. Then, click **'Nucleotide'** on the nucleotide-nucleotide BLAST page, so that you can start a new search.

Now, **search** again using our *Drosophila* tRNA, but choose the *E. coli* **database** option.

>gi|174308|gb|K00349.1|DROTRF2 D.melanogaster phe-trna-2
GCCGAAATAGCTCAGTTGGGAGAGCGTTAGACTGAAGATCTA
AAGGTCCCCGGTTCAATCCCGGGTTTCGGCACCA

Then, press **'BLAST!'** and then **'Format!'**

The output of our search is similar to the previous one—but note the E values. At this evolutionary distance (*Drosophila* to *E. coli*), there are only a few tRNAs retaining substantial similarity to the query sequence. The graphic display shows that most of these hits are towards the 5' end of our query sequence.

Click on the **'Score'** for some of the top subject sequences to see the alignments. Note that parts of our query sequence sometimes align with sequences within subjects that are thousands of nucleotides long. Remember that BLAST is a local alignment tool that finds high-scoring regions within a comparison. It does not usually produce an alignment from one end of the gene to the other.

The long subject file is for a long genomic sequence. There are 400 such segments in the database representing the entire *E. coli* genome.

Scroll down to look at the alignment for accession AE016756.1 (usually number 2 in the list) and remember the subject nucleotide position numbers for the alignment. Then click on the **GI** or **accession number** to access the Entrez data for this large genome sequence in *E. coli*.

Scroll down the data until you see subject nucleotide position for your alignment. Now click on the word 'gene' on the left of the position.

This link takes us to the Sequence feature view of the region. Scroll down to the bottom of the data, and you will see the actual sequence for this particular region.

BLAST does not attempt to do a global alignment from one end of the sequence to the other unless there is an uninterrupted high score throughout the alignment. Our hit reported a match of 23/25 for the 5'end of the molecule. If we align the complete sequence, we see that there are blocks of identity throughout the molecule with a match of 51/76 but only the block from 1-25 gives a high enough score to be reported by the BLAST program.

By examining the full sequence as shown in the following figure, it is obvious that the subject has significant similarity over its full length.

Remember the E value for this hit (0.015). This is the probability of finding a match of 23/25 in the database (*E. coli* genome) by chance alone. It is very high because the complexity of nucleic acids is low, having only the four possible base choices at each position. An alignment of two proteins with 23/25 identities would have a much lower E value because there are 20 potential choices at each position and the chance of finding the corresponding matches would be low.

```
5'
gccgaaatagctcagttgggagagcgttagactgaagatctaaaggtccccgg
gccgatatagctcagttggtagagcagcgcattcgtaatgcgaaggtcgtagg

ttcaatcccgggtttcggcacca
ttcgactcctattatcggcacca
                       3'
```

Now that we have had a look at highly conserved tRNA sequences, let's try mRNA. Messenger RNA sequences for the same conserved protein drift substantially through evolutionary time. In addition to the drift of the amino acid sequence over time, the degeneracy of the genetic code may have caused the same amino acid to be encoded in multiple ways.

Moreover, since the tertiary structural constraints are not as large on mRNA as they are on tRNA or ribosomal RNA, the sequences can diverge quite rapidly.

The p34 cyclin-dependent protein kinase is essential for mitosis and is strongly conserved at the amino acid level throughout the eukaryotic world.

Close the window with the *E. coli* results in it, and then click **'Nucleotide'** from the top of the previous search page. Now enter the accession number **'M12912'** (for the sequence for the p34 gene from *S. pombe*, the fission yeast) into the Search box, and click **'BLAST!'** and then **'Format!'**

When the results are in, we get a few strong subject hits. The graphic and hit list show strong alignments to several GenBank records for the fission yeast p34 gene itself. Not surprising, they are identical to each other. After that, however, the hits are only for very small regions, scattered throughout the gene and with significant scores in only a few other organisms, mostly other fungi. They do however show clustering in some particular regions presumably associated with conserved domains essential for function of protein kinases.

For comparison, if we run a BLASTP search using the protein product of the fission yeast p34 gene (accession P04551), we would get a very different result.

Close the window with the *S. pombe* nucleotide results in it, and then click **'Protein'** from the top of the previous search page.

Now enter the accession number **'P04551'** into the Search box, and click **'BLAST!'** and then **'Format!'**

Here the hit list is extensive with very low E values and clearly we have found a large number of homologues of this gene.

If we look at the third match (Click the score **'408'**), we see the homolog for *Homo sapiens*. Here you can see that the protein in the human genome is 66% identical and 80% positive (identities and similarities) to that of the yeast. Furthermore, the similarity extends over the full length of the protein.

There is a lesson in this. If you are searching for matches to a protein encoding sequence, then always search with the translated product.

Now that you have experience doing nucleotide and protein searches, do a search yourself using a gene that is mentioned somewhere in your textbook. Find the accession numbers for the DNA sequence and for the protein sequence in Entrez (make sure both are for the same species and that neither is for a multiple sequence file). Run BLASTN and then run BLASTP and analyze the result.

5. Learning to Use PubMed

Starting link: **http://www.ncbi.nlm.nih.gov.**
Note: For this tutorial Cookies must be allowed on your web browser. If you are unsure how to enable these, turn to your browser's help menu.

It is the responsibility of scientists to communicate their findings with their peers and also with the world at large. The major means of doing so is the publication of scientific papers and based upon these, textbooks. One of the most difficult challenges for the modern researcher is to keep up with the current research literature.

Researchers must relate appropriate publications to the problem at hand and use the findings of others to help direct their own progress. It is not an easy task. Even if someone had all day every day to sit and read, it would be a formidable accomplishment to say that they had read all relevant material within their own field, let alone that of others. The scientific literature is growing in size so rapidly that computer searches are indispensable. A second aspect of this is that access through the computer increasingly allows direct access to electronic publications.

You have probably already become familiar with searching the databases at your institution's library. In this tutorial we will take a closer look at the research literature database at NCBI. Within the Entrez program at NCBI is the PubMed database, which is produced in collaboration with the National Library of Medicine MEDLINE database. Most importantly, all of the publications at PubMed are cross-indexed to gene and protein sequences and all of the other databases at NCBI. They are also cross-referenced to textbooks in the form of key word searches. Notable in the list of textbooks is Griffiths *et al., Modern Genetic Analysis* (MGA).

Click **PubMed** on the top menu. We have already looked briefly at this page in previous tutorials. In our introductory tutorial we did a simple author name search for L. H. Hartwell.

Let's try a number of other searches to demonstrate the potential of this system. Let's start with the words in the title of Chapter 6 of MGA: 'Genetic Recombination in Eukaryotes.' First type **recombination** in the search window and press **Go**. The result is a daunting 130,000 plus research articles that mention the word in title, abstract, or key words. It is a popular and important area, but this is far too general a search term to be useful.

How about **genetic recombination**? Try it. We still get more than 110,000 returns. We are retrieving articles actually studying the process of recombination but also many that are of more peripheral interest to us, for instance, simply using recombination as a method to insert genes into chromosomes.

Try the whole title **genetic recombination in eukaryotes**. Now the output is cut to less than one thousand. That number would certainly be far fewer publications than are actually concerned with recombination in eukaryotes. The problem is probably with the term eukaryotes. The key words of publications might include phylogenetic group, but not be as broad as all eukaryotes.

Try **genetic recombination in mammals**. We get almost 50,000 entries. Many of these clearly do not have recombination as the primary subject of the paper.

Let's have a look at how our terms might be related in the formal subject headings within the National Library of Medicine. These Medical Subject Headings (MeSH) are used to classify publications. Click on **MeSH Browser** under PubMed Services on the left panel. At the new screen, type **recombination** in the search window and press **Go**.

Note the definition of the term and the display of various subheadings. This is more like the list of topics relevant to recombination that you would learn in a genetics course. Having a look at these terms sometimes helps narrow your search field by presenting you with some alternative (and narrower) terms. If we searched using the MeSH term *Recombination, Genetic*, we would get more than 100,000 returns. When we look at the subheadings covered we can see why this might be so.

For example, **transfection** is included. Click on it. The definition shows that it includes all papers that use transfection of DNA or plasmids into cells for experimental purposes. Notice that it is included in the techniques tree as well as the recombination tree. Many of the papers would be using it solely as a technique and the publication would not be about the process of recombination itself.

Click **Recombination, Genetic**, in the lower tree to go back to our original search.

Now let's try another term by clicking **Crossing Over (Genetics)**. This is narrower in scope and focuses on the chromosomal recombination process. Notice that it is listed under Cytogenetics in one tree, and under Recombination, Genetic in the other. Press the 'Add' button to open a PubMed search window for this term. Press **PubMed Search**. This yields a few thousand entries and a glance at the titles shows us that most are directly concerned with chromosomal recombination.

We could also focus our search by excluding some of the subheadings in the Recombination tree. Open the **MeSH browser** again (from the left-hand menu) and search **Genetic Recombination**. Click **Add** this term to the search. Now click **Gene Transfer, Horizontal** in the tree, and then toggle the drop-down menu to the right of Add to **NOT**, then click **Add**.

Next press **PubMed Search**. This returns an extensive list of papers dealing with various aspects of chromosomal recombination.

Now click the **Preview/Index** link in the bar under the search text box. This feature shows you a list of the searches you have run and the number of results for each. We could add more terms at this point. Toggle the drop-down menu at the bottom of the page to **Author** and type **Hartwell LH** in the text box. Press **'AND'**, then **'Preview'**. A new result appears in your list of searches with only a few entries. If you click on this link for the number of results, you will see a subset of Hartwell papers dealing with recombination processes.

There is a lesson here. It is relatively easy to define a narrowly focused search. For instance, we could search **recombination AND Hartwell LH** in PubMed and get a dozen or so journal references. It is often difficult however, to define a broad topic that does not include a substantial amount of irrelevant material (at least to you). If the search is narrowed to keep the amount of irrelevant material low, then it will almost invariably exclude material that you would want included in your list.

This means that you must use a somewhat broader search and then make a judicious choice of material to keep. Let's do a new PubMed search for **chromosomal recombination AND yeast**. If you wish to see the formal MeSH search terms for your search, press Details at the top. (To return to your search results, click on the number under the Result line.)

Our hit list now is down to less than a thousand articles, with the titles showing. To assess the relevance of each article, a simple click on each list of authors will give us the Abstract of the paper.

Move down the list and choose a title that you think might be relevant and click on its **authors** to display the Abstract. Go back and forth to find several that meet the criteria of being about chromosomal recombination in the yeast system. For each of these **check** the little box to the left of the reference. When you have found a set, go down to the 'Send to' button at the bottom of the page, **toggle** the drop-down menu to Clipboard, and then press the **button**. This will transfer these references to the Clipboard.

When you have finished, click **Clipboard**, on the gray menu bar. This will display only your selections and allow you to collect them by printing or saving to your computer. You could save it as a web page and retain all of the links to the abstracts. Alternatively, you could click Text to convert it to a text file that you can save to your disk and open with programs such as Notepad.

Now let's examine the links to textbooks. **Open** one of the abstracts on your list, and then click on the blue **Books** link at the upper right. This searches the keyword lists of the textbooks against the abstract. All words or phrases found will light up as a link. **Open** one. You are presented with a list of textbooks where the topic is covered.

Click on the link to the number of **items** in a specific book to go directly to the list of locations where your keyword is discussed.

Alternatively, you could open the books themselves by clicking on the **book cover**. (You might have to click the very small **Books** link on the top black bar.) Try opening the first edition of *Modern Genetic Analysis* to see the relevant table of contents. From here, you can you can search for a term in the search box to see the paragraphs that will help you define terms and understand the concepts. This will allow you to read many scientific abstracts in areas that where you have little experience as yet.

From the book's table of contents you can also **open** the chapters directly. You will see various terms underlined in the text. **Click** on one of these. A glossary appears to provide you with a precise definition.

Initiate your own search by investigating Holliday structures in recombination. You might narrow the search by including a term such as **'sequence'** or perhaps **'genes'**. Use the resources of Entrez to find the sequence of at least one of the proteins involved in resolving the complex.

6. OMIM and Huntington Disease

Starting link: **http://www.ncbi.nlm.nih.gov.**

We looked briefly at the Online Mendelian Inheritance in Man (OMIM) resource in the introductory tutorial. Now we will explore it more fully. Note that if you are searching for information concerning a particular gene or disease, an OMIM search can be run directly from the initial drop-down menu on the NCBI home page. This is done by choosing OMIM and entering a search term.

However, in this tutorial we want to look at some of the structure of the OMIM database and this information is available from the OMIM home page. Click on the **OMIM** link on the NCBI home page to get to the OMIM home page.

Now let's start with the OMIM statistics page that we saw earlier. Choose **'Statistics'** from under the heading 'OMIM Facts', on the list on the left.

The statistics page shows the number of loci included in the OMIM database, as well as a breakdown of their distribution by inheritance pattern. Remember that our most recent estimate of the number of human genes is about 35,000. Now, notice that barely one third of all human loci are currently included in this database. There is still a long way to go before they are all categorized.

The various links in the table will call up a list of the genes fulfilling a particular mode of inheritance. Click the first **number** in the X-linked column.

The link takes us to a list of genes in the format of an Entrez search result. The individual records, however, are to annotations within the OMIM database. The Entrez links to sequence and related genes are over on the right side of the page. Let's look at a typical OMIM record by clicking on the first record **'*314998'**.

This record is a description of the discovery of the gene. It includes the literature references and information about linkage and location. This is all within a brief description written by a curator. This is an actively annotated database that changes with time. As research progresses, records of this type must be updated to keep them current.

Now, click on the green **OMIM logo** at the top of the page to get back to the OMIM homepage.

From here, click on '**Update Log**' from under the heading 'OMIM Facts', on the list on the left to call up a table of the frequency of updates and new additions of records to the database. Note that approximately 100 new records are added each month. Many times that number of existing records are updated to reflect our increasing knowledge of particular genes. Now, click on the green **OMIM logo** at the top of the page to get back to the OMIM homepage.

Now let's look at a particular case, Huntington disease. Type '**Huntington**' into the OMIM search window, and press '**Go**' to see the display for the OMIM record for this gene. The complete record for such an important and well-studied locus is tens of pages long. We will examine a few of the features.

First, click on the '**Links**' button on the right of the first entry (accession number *143100), and then choose '**Protein**' to take us to the Entrez protein record for Huntington.

Next, find the entry for accession number NP_002102 by doing a **Protein search**. Then, click on the accession number '**NP_002102**' to see the record.

Scroll down through the human huntingtin protein record until you reach the Comment section. Notice that here is a description of the protein and its expression, the phenotype of mutants and even the nature of the common mutation in Huntington disease. The disease is associated with the expansion of a trinucleotide repeat sequence leading to an enlarged polyglutamine stretch in the protein. Notice also in the text how the NCBI staff indicate precisely the history of this record relative to the initial submissions to GenBank.

From here, scroll to the top of the page and click on the blue word '**Links**' in the cluster of links to the right of the heading for huntingtin. From the drop-down menu that appears, choose '**OMIM**'. Next, click on '*143100' to reach the OMIM entry for this disease.

The OMIM record for Huntington disease is a very rich source of information. The research literature is summarized under various headings including Clinical and Biochemical Features, Inheritance

Pattern, Population Genetics of the disease, and lots more. Clicking on the various headings on the left will take you to the subsections of the file. This is an immense resource. Entire term papers could be written based on this page alone! Clicking on the Mini-MIM button will take you to a much simpler version of the page. Clicking on any of the light bulb icons gives you the relevant literature sources.

Now let's have a look at the gene map. Click on the **'Gene Map'** button on the left.

The map button takes us to a portion of the gene list for chromosome 4 and displays the 4p16.3 region, which includes the Huntington locus, and is listed here by the name of the gene product, huntingtin. Click on **'4p16.3'**.

Now we have the graphical display of chromosome 4 with the Huntington disease locus highlighted. We can see from the graphic of the entire chromosome on the very left that the locus is very near the telomere on the *p*-arm of the chromosome. (Remember that the two arms of a human chromosome are named *p* and *q*.)

By zooming in on the locus (use the zoom in feature just above the chromosome graphic on the very left), we can display the genes that are located close to the Huntington locus. The list on the right is compiled from sequence data. Under the Morbid column are various loci associated with disease phenotypes.

Now, click on **'HD'** under the "symb" column to go to the LocusLink record for the Huntington disease gene.

LocusLink is another annotated database and a very rich resource. Scroll through it. As with all of the other resources we are looking at, it is richly decorated with links to databases and information resources of all types. In the colored boxes are links to PubMed, OMIM, and other databases we have touched upon. However, some are new. Let's click on **'HGMD'**, the human genetic diseases database.

The HGMD database is an outside resource specializing in the nature of characterized mutations for various loci. Note that only the trinucleotide repeat is listed as present for the HD gene.

Now, let's go back to the Huntington disease LocusLink page at **http://www.ncbi.nlm.nih.gov/LocusLink/LocRpt.cgi?l=3064**.

This page will take us back to the gene map page if we click on the yellow **'MAP'** box in the menu.

Now we are back at the map view of the Huntington region. We can go to greater and greater depth from this map, which is largely a linkage map, to the actual nucleotide level. There we can examine the structure of the gene itself. Click on **'sv'** in the highlighted huntingtin row (the row with 'HD' in the 'symb' column – you may have to scroll down the page slightly).

'SV' stands for sequence viewer. In this case, it links to a graphical display of the annotated nucleotide sequence for the huntingtin region on chromosome 4. If you clicked on it for other genes in the display, you would activate different regions of the chromosome for viewing. The display shows about a megabase of DNA with several open reading frames indicated graphically. An expanded view of a particular region can be seen by clicking on a particular region of the chromosome and zooming in or out using the tool on the right.

The expanded region displays the start of the huntingtin coding region, indicated by HD at the left, and the three parallel lines under the sequence. Below this is the DNA sequence itself.

Scroll further down along in the sequence and you will see the transcription start site for the huntingtin mRNA. A bit further along, you can see the start of the coding region for the huntingtin protein with the amino acid sequence indicated below. Note the QQQQQ polyglutamine region — this is the region that expands and interferes with function, and is the mutational event that causes Huntington disease. The forward and backward blue arrows at the top of the sequence allow you to walk in both directions along the DNA strand at will.

We have explored OMIM to a sufficient depth for the moment. It is clear that there is much more here. As with all of the other sections of these tutorials you must go to NCBI and examine the links yourself.

Choose a well-defined human genetic disease based on one discussed in your textbook or class. Use OMIM to identify its chromosomal location and determine the nature of the gene product. What is the biochemical lesion involved? Use sequence viewer to find the start and stop sites for the gene. Also find the 5'-end of the first intron in the gene.

7. Finding Conserved Domains

Starting link: **http://www.ncbi.nlm.nih.gov**.

In the Protein BLAST tutorial we saw how the BLAST site at NCBI first compares the query sequence to a database of conserved patterns or motifs, the Conserved Domains Database. The patterns in the database are designed to detect structural motifs in proteins. By focusing on these conserved domains, such searches help us assign functions to unknown genes and see the domain architecture of complex proteins. We have already seen output from such a search for the dystrophin protein. Let's start there to investigate this feature more thoroughly.

Instead of starting with the BLAST page (where a Conserved Domains search is run as part of the BLAST search), let's go directly to the Conserved Domain Search page.

Click on **Structure** on the top black menu bar, then on **CDD** (for Conserved Domains Database) on the left-hand side bar. (You will have to scroll down to see it.)

Our basic evolutionary model is that sequences will drift further and further apart over time unless selection is at work. When we do a BLAST search, we compare proteins looking for regions of sequence conservation. When we find such evolutionarily conserved regions, we assume that they represent amino acid sequences that are responsible for the compact functional structural features of the proteins.

The order of the various conserved features that might be present in a protein is referred to as its 'domain architecture.' Complex domain architecture is very common among proteins in the human genome and other similarly complex organisms.

The CDD page contains a search entry window as well as information about the program and its underlying databases. Note also that it is closely integrated with the structural databases MMDB, Cn3D, and VAST, which are accessible from the side panel. We will learn to use these in a subsequent tutorial. Keep in mind for the moment that a conserved domain is really a compact structural feature, perhaps a binding site for another protein, or a catalytic pocket.

First, let's look at the nature of a search pattern for a conserved domain. Click on **Pfam**, in the paragraph beneath the search window. This takes us to Washington University in St. Louis, the home of the Pfam (Protein Families) database. Each Pfam search pattern is derived from a multiple alignment of a family of related proteins. The pattern is designed to recognize members of that family when an unknown sequence is searched against the Pfam database.

In the list of options, click on **BROWSE PFAM**.

Pfam is a large database with over 3000 conserved search patterns for all sorts of different conserved domains. The page that first appears shows the top twenty families, a sort of popularity list. Other lists are organized alphabetically. Each line in the top twenty families list gives its name, an accession number within the database, some statistical information, a link to its 3D structure, and a description. This database is an extremely rich source of information about proteins.

Let's try one. Click on the **LRR** motif, the Leucine Rich Repeat. An information page should appear telling you about the motif and what its function is thought to be—a protein-protein interaction domain. If you scroll down the page, you can retrieve the protein alignment that gives rise to the conserved pattern. Click on **Retrieve alignment**.

This list is hundreds of entries long. Note that sequences from the various genes (accession numbers on the left) are all a little different, but similar to each other. Also note that the pattern is not a continuous stretch of amino acids but instead is most highly conserved in three distinct clusters. In generating the Pfam search profile, the database combines all of this information to generate a pattern that will recognize all of these entries and others like them. Click the **back** button on your browser to go back to the information page.

Now try the **Retrieve domain structures** button to the right of the Retrieve alignment button. Here we can see the domain architecture for a number of proteins with leucine rich repeats (in green). Notice that different proteins have different numbers of domains and they are located in different locations along the primary sequence. What they all have in common is a conserved primary amino acid sequence in that region, which reflects a compact conserved structural domain with similar 3D structure in each case.

This is a diverse group of proteins. They also have a variety of other functional domains. If you roll your mouse over the domains you can see what they are.

Let's go back to the NCBI CDD page and run a search. Go to **http://www.ncbi.nlm.nih.gov/Structure/cdd/cdd.shtml**.

When we run a CD-search, our query sequence is compared to the thousands of stored patterns located in the Pfam database for various protein families (and also the Smart database at EMBL, a related resource).

Let's run a familiar search. As we have already seen, dystrophin has a complex modular structure containing a number of conserved functional domains. Type the accession number for *Drosophila* dystrophin (**XP_081212**) into the Search window in the Run CD-Search box in the middle of the page. Leave the Search Database drop-down menu toggled on **CDD**. (If you open this menu you will see that you can restrict the search to just Pfam or just Smart.) Now press **Submit Query**.

The graphic displays the location of a variety of conserved domains located along the length of the dystrophin protein. It is similar to the result we saw when we ran human dystrophin in BLAST.

Move your **mouse over** the colored domains and watch the display window provide information about the different regions.

Below the graphic is a list of the domains found. It tells you the Pfam or Smart accession number, a verbal description, and the E value for the match. Lastly, there is a detailed alignment for each of the domains found as well as considerable descriptive information about them.

As you can see, with a single search, researchers are able to glean an enormous amount of information regarding the structure and potential function of various regions in the protein.

Let's go back to the CDD search page by clicking the **back** button on your browser.

By analyzing dystrophin for conserved domains, we have discovered its domain architecture, that is, the order of different domains along its primary sequence. This was visualized as the graphic display with the various features arrayed along it. In principle, functional homologs of our protein would have a similar set of domains arrayed along them. When we did our BLAST searches for similarity, we found proteins with similar primary sequences to our query. In the CDART program (Conserved Domain Architecture Retrieval Tool) we are able to search for a pattern of conserved domains rather than for the primary sequence itself. Let's try it.

Click on **CDART** at the left side of the screen. Enter our dystrophin accession number (**XP_081212**) in the search window and press **Search**.

In the result, the graphic of our *Drosophila* dystrophin domain architecture is at the top. Below it are proteins with similar domain architectures. If you click on any of the icons on the graphics, it will open a window with the Pfam or Smart pattern information page.

Notice in the graphics aligned below our gene that some of the domains are in similar positions to those in our protein, notably the two calponin domains to the left, and the WW and ZZ domains on the right. The spectrin repeats are more variable, but present. In addition there are other domains that are not present in *Drosophila* dystrophin. This program is scoring similar domain architecture patterns and is a very rich source of information.

Note that the overall output of our search is about 20 pages long. **Click** on some of the later pages at the bottom and see some of the more divergent comparisons.

Remember that the Pfam and Smart motifs are identified by running programs such as BLAST to find families of related sequences. The patterns are then distilled from the conserved regions of these gene families. These patterns can now be used to search and identify new sequences and to place them in functional families.

Choose a gene that interests you, either from your text or related to your earlier searches, and run a CD and DART search.

8. Determining Protein Structure

Starting link: **http://www.ncbi.nlm.nih.gov**.

WARNING: Depending on how fast your computer is and how much memory you have available, your computer may be very slow in responding to parts of this tutorial. To speed it up, we suggest closing any other programs you may have open before you start the tutorial.

Genome sequencing has given us an enormous quantity of data regarding the genetic breadth and potential of living organisms. As we have seen, tools such as BLAST make it relatively easy to search the genome databases using the primary sequence of hypothetical proteins. This allows us to ask questions about similar proteins, homologs, and gene families. The primary sequence determines the three-dimensional shape of a nucleic acid or protein. Structural databases such as Entrez Structure at NCBI contain details of protein structure based on X-ray crystallographic and NMR studies. From this database it is possible to retrieve and examine a particular three-dimensional structure in a viewer called Cn3D. In this tutorial, we will examine some structures and learn to use the Cn3D structural viewer.

First, let's download the Cn3D viewer plug-in for your browser. Click on **Structure** at the top black bar of the NCBI homepage. Then, click on **Cn3D v4.0** on the left-hand side menu.

Cn3D provides a graphical view of the three-dimensional structure for a molecule. Follow the directions to download the program. First click on the appropriate operating system for the computer you are using (**PC, Mac, Unix**) on the blue bar at the top of the page. Read the system requirements and directions carefully. In some cases (for PCs) there is more than one version of the program. Next, click on **here** to start the process. Either allow the file to open automatically, or save it to your disk and then click on it to run. The program will install itself and will recognize structure files automatically when we later download them.

Now let's go back to the Structure page by clicking the **Structure** heading at the top of the page.

The structure database can be searched through Entrez. In the search text window, type **1JM6**. This is the identification number for rat pyruvate dehydrogenase in the PDB protein database. Click **Go**. When the result appears, click **1JM6** to see the NCBI Molecular Modeling Database (MMDB) Structure Summary page.

This page provides a description of the file including the MMDB ID number, PubMed link, species, and citation information. Make sure that you downloaded the viewer, then click on **View 3D Structure**.

You will see the file download and then two popup windows for Cn3D will appear. Move them to the left if need be so that you can still follow this text.

The upper pop-up window (Cn3D) shows a representation of the 3D structure of the protein. This is drawn from the 3D crystal coordinates for each of the atoms in the molecules. The representation is in structural mode: alpha helices are shown as green cylinders and beta pleated sheets as gold flat arrows. The connecting amino acid chain is in blue.

The display is interactive. The three scroll bars move the molecule in three dimensions. Click on the right hand **arrow** on the bottom scroll bar and hold down the mouse to rotate the molecule in the horizontal plane. Try it. The calculations are done on your computer and the rotation speed will be affected by how fast your computer is. As it rotates, note that it consists of two mirror image complexes consisting of the A and B chains. Any view of the molecule may be obtained.

As the structure rotated, did you notice the bound ADP molecule in each complex (partly colored in red and complexed with a Mg ion)? Can you identify the adenine ring structure in the ADP molecule? You may enlarge this window to full screen to make it easier to see, but when finished, reduce it again. Alternatively there is a Zoom In feature under View in the top menu.

The lower pop-up window (Sequence/Alignment Viewer) shows the primary sequence of the two amino acid chains. Marked along its length is a representation of the helical and pleated sheet domains. Notice that the primary sequence is color-coded to match the representations in the 3D viewer. To see the other chain, open **File**, then **Sequence list**, and choose the other one from the pop-up menu.

Now go back to the Cn3D viewer window. The representation is of secondary structure. On the menu at the top, open the **Style** drop-down menu, open **Rendering Shortcuts**, and then click on **WireFrame** to see a different representation of the structure of the amino acids chain. It is still color-coded for the different structural features. Lastly, try **Spacefill** under Style. (This model requires much more computation for each view and may take several minutes to download. It will also be slow when you rotate it. If your computer is very slow, you should skip this step.) This is a more realistic view of what the surface might really look like, if you were another molecule! Try rotating it.

Close the pop-up windows and let's look at another example. Click **Structure** at the top of the page to go back to the homepage.

The Structure database is linked to Entrez. Therefore we can enter terms and search in the same manner that we have done in other tutorials. Enter '**GCN4 AND yeast**'. This protein is a transcription factor that binds DNA. Click **Go**.

The list of hits is two pages long and represents all of the entries in PDB for this protein, its fragments and various mutants. As with all of the databases that we have looked at, it takes a certain amount of sorting to get what we want. Go to the bottom of the list. (Select page: **2** in upper right corner then scroll down the page). Choose the last entry, **1YSA**, by clicking on it. The MMDB structural summary for 1YSA is shown. Click on the **View 3D Structure** button. As before, the two Cn3D pop-up windows should appear.

This view shows us the interaction of chains of GCN4 transcription factor (in green) bound to a promoter site on DNA (in red). We are initially looking down the length of the DNA helix. First, let's switch the view to **WireFrame** in the Style menu at top. This will let us see the DNA base pairing. Now **rotate** the image through about 90 degrees using the lower scroll bar. Stop when the red/blue nucleic acid chain is vertical. Now **rotate** the molecule using the lower scroll bar. As it rotates we can plainly see the DNA double helix with its planar base pairs (in blue). We can also see the GCN4 chains contacting the DNA strands on each side. This is the sequence-specific recognition of the DNA molecule by the transcription factor. We can also see that the protein chains interact with each other at their opposite ends.

Now take a look at the Sequence/Alignment Viewer. It can be toggled (under File, sequences) to show either the protein or nucleic acid sequence. If you use your mouse to highlight a part of the nucleic acid or protein sequence (whichever you have showing in the panel) then it will color the corresponding sequence in the 3D graphic. **Try it.**

Lastly, place the viewer in **Spacefill** mode to have a more realistic sense of what the molecules are really like. (Again, depending on the speed of your computer, this may take some time, though not as much as the first protein, as this is a much smaller molecule.) **Close** the viewer windows, and go back the Structure homepage by clicking **Structure** at the top of the page.

In addition to allowing an investigator to search for the known structure of a particular protein, NCBI also allows an investigator to search the database for structures similar to the one of interest. This allows a crystallographer with a newly acquired sequence to search for structurally related proteins utilizing the VAST program (Vector Alignment Search Tool).

Within VAST, all of the structures in the MMDB database have already been compared to all other structures in the database. Since we don't have a new crystal structure to try, let's explore the structural comparison function using a previously known sequence. Type the PDB identification number **1G5S** in the search window and click **Go**. This displays a record for human cyclin dependent protein kinase, one of the proteins that controls the cell cycle.

First let's click on **View 3D Structure** to see what this protein looks like. **Close** the OneD-Viewer window, and minimize the Cn3D viewer window for the moment.

The Structure Summary page includes a line labeled 3D domains 1, 2, 3. These symbols refer to the subdomains of the protein. These links take us to the most structurally similar files in the database. Click on **2**.

A VAST Structure Neighbors summary page appears. Notice that it has a table of records with varying degrees of amino acid identity. Click on the small **square** to the left of 1IA8, near the top. This record is for another protein kinase found in humans. Now click on **View 3D Alignments**. A new Cn3D window will open. This contains a comparison of 1IA8 to our original 1G5S file.

Enlarge the original Cn3D window containing 1G5S and arrange it along side of the new window containing 1IA8. Keep the sequence alignment window for 1IA8 open below it. Toggle both windows to **Worm** under **Style**, Rendering Shortcuts. Notice the overall similarity of shape and organization to our first protein kinase. Note that in the new window the full structure of 1G5S is shown, but only the second domain of 1IA8. This is not surprising since they are both protein kinases and we have found the second one by structural comparison to the first. **Close** the 1G5S window.

A careful look shows that the new window contains both the 1G5S and the 1IA8 structures superimposed on each other. You can highlight the two sequences by using your mouse to highlight portions of the sequence in the alignment window DDV at the bottom. For instance, **highlight** amino acids 110 to 130 on the 1G5S (top) sequence. (There is a position counter in the lower left corner of the window.) You can see that one of the alpha helices in the 3D model turns yellow. If you look carefully you will see that close to the yellow is a red and blue chain directly paralleling the yellow one. This is the 1IA8 sequence. If you now use your mouse to **highlight** 110 to 130 on the 1IA8 sequence, you will see the corresponding change on the 3D image. The VAST program statistically compares how closely the two models fit to each other and aligns them in three dimensions.

Programs such as VAST are capable of finding very distant homologs, even where we do not see statistically significant sequence identity using alignment programs such as BLAST. Ultimately, it is the 3D structure of a protein that is selected during evolution. It is possible to have similar 3D structures, but quite different amino acid sequences. **Close** all of the pop up windows.

This is just a brief look at the richness of the 3D information that is available. Such information is critical to researchers wanting to understand how proteins interact or how a drug inhibits a protein.

Go back to the NCBI Structure page and search in Entrez Structure for a structure that interests you. It could be one of the sequences used in earlier exercises or something new. Keep in mind that only a small proportion of all sequences in GenBank have had their structure determined. As a start, you might try looking up the structure of a transfer RNA.

9. Exploring the Cancer Genome Anatomy Project

Starting link: **http://www.ncbi.nlm.nih.gov**.

A number of specialized databases are available through NCBI. These often represent collaborative efforts between outside resources and NCBI. We have already made use of several of these, for instance in the Conserved Domains tutorial and in using OMIM.

Many others are listed on the right hand side of the NCBI homepage. The Cancer Genome Anatomy Project is organized by the National Cancer Institute. NCBI provides a number of database and bioinformatic resources to the project. This allows cross-linking among all of the related databases.

Cancer is a very complex disease. In general, it is characterized by inappropriately controlled cell proliferation. The resulting cells interfere with other tissues and they characteristically spread in the body. There are many ways in which a cell's machinery can go awry and still have this overall end result. Cancer, therefore, can be thought of as hundreds of different diseases.

Most cancers arise from spontaneous somatic mutations accumulating over your lifetime. In this sense it is a disease of aging, although it too frequently occurs in younger people. Somatic mutations, of course, are not passed on to the next generation. There are also cancers caused by inheritable germ line mutations and they can be followed through families by pedigree analysis. Characterizing all of the genes and mutations that directly cause or predispose us to cancer is the goal of the Cancer Genome Anatomy Project.

Click on **Cancer Genome Anatomy Project** in the right-hand menu. This page contains a description of various aspects of the NCBI-CGAP collaboration.

The Cancer Genome Anatomy Project (CGAP) pulls together a large amount of public data on genes altered in cancer cells. This ranges from gene sequence, mutational analysis, cytogenetic localization, chromosomal aberration, biochemistry, and clinical manifestation for cancers. By making all of this information easily accessible, CGAP helps the researcher analyze new information and relate it to what is already known. Since cancer ultimately is a problem of expression of a mutated gene product(s) and consequent inappropriate expression of other genes to give the cancer phenotype, analysis of gene expression has been one of the major efforts of CGAP.

Click on the **Cancer Genome Anatomy Project** link.

This is the homepage for CGAP (**http://cgap.nci.nih.gov/**). Notice that we can enter the databases from several directions and that they are extensively linked with NCBI at various levels. We could ask about gene or tissue expression patterns, chromosome aberrations, biochemical pathways and various other resources.

Let's try several. First, click on **Genes**. The GENES page provides access to a number of bioinformatics tools. Click **Gene Finder** in the top section.

Gene Finder allows us to search the database by gene name, accession number, or tissue where a gene is expressed. The key words and identifiers are from the UniGene classification at NCBI. Let's try an example. In the lower search window, toggle the Tissue Type to **Breast/Mammary Gland** and then type **BRCA1** in the Gene Name text box. BRCA1 (breast cancer associated) is known to be involved in some forms of early onset breast cancer and occurs in families as a germ line mutation. Press **Submit Query**.

Gene Finder returns the BRCA1 gene and a number of records for it. It also returns records for a number of genes whose gene products are known to interact with the BRCA1 protein. The **Gene Info** link associated with BRCA1 takes us to a summary page containing the various GenBank accession numbers as well as links into various databases. Click on it.

The Gene Info page is a compilation of information and links concerning our gene. Scroll down the page and look at the scope. Near the top are a number of database links. Click on **UniGene**. The pop-up window that appears gives a summary of information about the gene and about the various GenBank records for it. Note the related proteins in a variety of eukaryotes, even plants. Remember that cancer is a disease that affects the basic cell proliferation machinery and much of this is conserved in all eukaryotes.

Close the pop-up window.

Clicking **LocusLink** takes us to a wealth of information about BRCA1 in the LocusLink database at NCBI. Try it. Scroll down through the page. **Close** the window when you are finished.

Similarly, clicking **OMIM** takes us to a long list of entries on the nature of the gene and its inheritance. We have looked at some of these resources before. It would take many hours to follow all of the links and text. **Close** the window.

Remember that these databases could also have been found by entering BRCA1 in Entrez.

Now let's go back to CGAP and see what other types of information it might offer. Click on the **CGAP icon** at the top left of the page.

As we unravel the biochemical function of various genes and gene products, we can place them in functional pathways using all of the information available to us. Click on the **Pathways** link.

Click on the **BioCarta Pathways** link. BioCarta has an extensive list of biochemical pathway charts. Under A, click on **ATM Signaling Pathway**.

Notice the schematic of the cell cycle across the top of the map. The ATM pathway is important for delaying cell cycle progression to allow time for repair if there has been damage to DNA. This helps to prevent permanent damage or chromosome breakage and the consequent genomic instability. Maps of this type help the researcher place a particular gene product into a broader context. Our BRCA1 gene product is shown just to the left of center. Its precise role, since mutations in BRCA1 lead to a high risk of breast cancer, is the focus of much research around the world. Clicking on any of the icons in the map takes us back to the Gene Info page for that gene. You might have noticed, when we examined it, that the bottom of the BRCA1 Gene Info page contained a link to this map.

Let's go back to the CGAP Genes page and try another link. Click on **Genes** in the green menu near the top of the page.

One of the problems facing all researchers (and students) is the development of a formal language describing the function of genes. This is essential if software is going to be developed that will allow us to move transparently among a variety of different databases. An 'ontology' defines function in a hierarchical fashion, moving from the general to the specific. Click on the **Gene Ontology** (GO) browser.

The pop-up page that appears contains a very general opening page with three descriptors: Biological Process, Cellular Component, and Molecular Function. The + sign in the boxes indicates that there are hidden trees for each.

Click on each of the + **signs** next to the three terms in turn and open up the next level of the hierarchy. Each line starts with a symbol.

Try clicking on the + sign next to 'binding', and then 'nucleic acid binding' under Molecular Function. This opens up a further four categories including whether gene products bind DNA or RNA. (You may have to scroll down the page to find it.)

Click on the + **sign** next to 'DNA binding' and a further 20 categories open with increasing levels of specificity.

If you click on one of the numbers it will call up a list of the genes in the main browser window. This pop-up window will go to the background and you will have to click on it in your tool bar to bring it to the foreground again.

In some ways working through this hierarchy is the opposite of what we have done in previous tutorials. We have started with a gene and built our information tree from the gene up, until we get clues to function. In the hierarchical structure of the Gene Ontology we narrow the functional fields and then ask "What genes belong to this group?". These terms and the hierarchical structure play a major role in categorizing the genes of organisms as new genomes are sequenced. The terms provide a common language for comparison. Keep in mind that BLAST scores are a major criterion in deciding whether a newly discovered sequence has a similar function (and ontology) to a protein whose function has been previously investigated.

A gene product may belong to more than one category. This is especially true of human genes where very complex domain architecture often correlates with multiple functions.

Let's go back to the CGAP/Gene home page. **Close** the Gene Ontology Browser, and click on **Genes** in the green menu near the top of the page.

We now know of several million different single nucleotide polymorphisms (SNPs) in the human genome. In the case of hereditable disease, such as BRCA1 associated breast cancer, we can track the alleles responsible. Most importantly we can use the information for diagnostic tests to assess the risk to individuals. Click on the **SNP Gene Viewer** in the second section.

Again, a pop-up window will appear. Type **BRCA1** into the search window and press **Submit**. A list of SNPs for the BRCA1 gene and some BRCA1 associated proteins appears. Notice in the column at the right that many SNPs do not cause changes in amino acid sequence in the gene product. Click on **view** under the accession number of one of the BRCA1 mutations that does cause a change. A graphic appears at the bottom of the page along with a list of SNPs and the amino acid change (e.g. Pro871Leu is a change from proline to leucine at amino acid position 871). The graphic of the gene also displays the conserved domains for the protein to orient the viewer. Each point mutation is a different allele.

Let's **close** the pop-up window and go back to the CGAP main page and examine other resources by clicking the **CGAP icon** at the top left of the page.

Next, let's try clicking the **Chromosomes** link. Several resources are listed. Click on **GeneMap99** in the blue menu at the left. A new pop-up window will appear.

The GeneMap project compiles mapping data for the human genome. It maps known genes as well as ESTs. ESTs, or expressed sequence tags, are fragments of cDNAs derived from mRNAs in the cell and thus are a part of a gene sequence. ESTs represent a major portion of the sequences submitted to archival databases and at NCBI they are held in a database called dbEST. Let's look at what has been mapped.

Click on **Chromosome 1** on the upper menu. This yields a graphic of chromosome 1 as well as information on what has been mapped. Notice the red tag on the left hand vertical line. If you move your mouse over this line and click on various regions, you will see the red marker move. As this moves to new regions, the list at the bottom of the page (scroll down) shows you the physical mapping information for all known genes in the region as well as various ESTs that have been mapped. Such detailed mapping information is instrumental in compiling and interpreting the human genome sequence. It is of tremendous value in detailing the nature of chromosomal rearrangements that are often common in cancer cells.

Now let's **close** this pop-up window and go back to the CGAP homepage by clicking the **CGAP icon** on the top left of the page.

CGAP aids the researcher both by being an archive of information as well as being actively engaged in the compilation of new information and resources. Start your own search through the system. Perhaps begin at the BioCarta map for Cell Cycle: G1/S checkpoint. Checkpoints delay cell cycle progress while DNA repair occurs and many of the proteins in these pathways are associated with cancer. Find the p53 gene and click on it. This gene is mutated at high frequency in human cancers and you will find a wealth of information about its function.

10. Measuring Phylogenetic Distance

Starting link: **http://www.ncbi.nih.gov**.

Natural selection depends upon heritable variation within a population. The variation is a result of random mutation as well as the random choice of gametes. The net result is phenotypic differences upon which natural selection can work. Differential reproductive rates based on these phenotypic differences result in changes in the genotypes represented within the population over time.

It follows that over long periods of time, random change and selection will lead to increasing DNA sequence divergence among organisms. With certain underlying assumptions, we can use this information to estimate evolutionary distance. We have already looked at this to some extent in the BLAST tutorial and also when considering the nature of Conserved Domains. In this tutorial, we want to explore the relationship in a bit more detail and to examine some of the available resources.

NCBI is the repository for vast amounts of sequence data. It currently has at least one bit of sequence from each of more than 100,000 different organisms. By comparing homologous genes among different organisms we can get a measure of sequence divergence and from that, relative estimates of evolutionary relationship. Different genes, however, may diverge at different rates, so evolutionary relationships inferred from limited data must always be treated with a certain amount of skepticism. Clearly these data must also be combined with insight gained from comparisons of phenotypic characters and the geological record to reach a consensus within the scientific community regarding the relationship of one organism to another.

First let's look at the taxonomy resources at NCBI. NCBI has undertaken to generate a consensus taxonomy linked to all of the sequences in the database. Obviously sequence data plays a major role in defining such relationships. First let's look at the NCBI taxonomy resource. Click **TaxBrowser** on the upper menu.

This page is the gateway to a variety of resources. One of the goals of the Taxonomy section at NCBI is to generate a consensus Taxonomy for all of the organisms represented in the database. Let's have a look at one of our most commonly used organisms. Click on the **Drosophila melanogaster** link in the organism list.

This opens a summary page of taxonomic position as well as a list of the number of records of various types in the database. The taxonomic tree for *Drosophila melanogaster* is shown. Click on the word **Drosophila** to open the Genus level of the taxonomic tree.

There are a lot of fruit fly species in the world! On the upper gray menu, **toggle** the **proteins** to on and press **Display**. This will display the number of sequence records available for each item. Notice that for some, there is only a single entry.

Search down the list and find the melanogaster group (under Sophophora), melanogaster subgroup and eventually *D. melanogaster* itself. Each of these species has a unique genome. Our premise in looking at sequence data is that it would diverge more and more relative to *D. melanogaster* itself as we went further up this taxonomic tree. The quality of our comparison is dictated by the amount of sequence available for each species and the extent of our comparisons. Although we have the entire genome for *D. melanogaster*, that is not the case for the others.

To use sequences in order to help build a phylogenetic tree, we must align the sequences with one another to score their divergence.

In the BLAST tutorial, we saw that more distantly related organisms had more divergent sequences. When we examined the output from a BLAST search, the hits were ordered from most similar to most divergent and ranked by the BLAST score.

BLAST results are a complicated list of accession numbers and formal species names. It is often difficult to understand the relationship of the various organisms to each other simply by looking at the list. Many times, without expanding each entry, we cannot tell what organism we are looking at.

BLAST results, however, are linked to the Taxonomy browser. Let's run a BLAST search and explore it. Click on the **NCBI icon**, then **BLAST** and **Protein BLAST**. We will search using an *Arabidopsis* PCNA sequence. This protein is part of the DNA synthesis machinery and is strongly conserved in all organisms. Enter the accession number **NP_180517** in the window, press **BLAST!** and then press **Format!.**

As we have seen before, the BLAST results page is a detailed list of hits, but with limited visible information about the organisms in the list.

Clearly we can open each entry and examine them one by one. However, BLAST results are linked to the Taxonomy Browser. Go to the left of the page (above the graphic display of results) and click on **Taxonomy Reports**.

The BLAST result is now formatted in the new window with the lineage of *Arabidopsis*, our query organism at the top, and then each hit on the list arrayed according to phylogenetic position.

Clicking on the **Organism Report** link at the top takes us down the page to list the hits from each organism.

Similarly clicking **Taxonomy Report** gives us the consensus taxonomy with all of the hits listed in it.

This is a very useful tool for orienting yourself to the strange organisms that we deal with.

Close the pop-up windows.

It is possible to use the quantitative measure of sequence divergence to construct phylogenetic trees. The underlying assumption is that the sequence divergence is related to the evolutionary distance.

We will build a small phylogenetic tree based on a sequence comparison of a single protein that is strongly conserved in all organisms. The protein is PCNA, which we used above, and which is part of the DNA replication machinery. We will use the ClustalW program to generate a multiple sequence alignment and then the Phylip program to form our tree. These two programs are not available at NCBI, so we will run them at the European Molecular Biology Laboratory (EMBL) site, one of the archival sequence databases centers in Europe.

We have looked at sequence file formats in earlier tutorials. For this exercise we need to construct an input file for the ClustalW program. We will do this as a multiple FASTA file of PCNA sequences chosen from a number of different organisms. This can be done by collecting several different PCNA sequences on the Clipboard at NCBI, displaying them in FASTA format and saving the result as a text file (.txt) that can be edited in Notepad or a similar text editor.

Let's choose some sequences. Go to the **NCBI** homepage by clicking the icon at the top left of the page. Then, go to **ENTREZ**, toggle the search to **Protein** and enter **pcna AND human** in the search window. Press **Go**. In the output list, **check** the box in front of accession number P12004, and then at the bottom of the page, toggle 'Send to' to **Clipboard** and press **Send**. **Repeat** the search for *Arabidopsis thaliana* (AAC95182), *Drosophila melanogaster* (NP_476905), *Danio rerio* (NP_571479) and *Gallus gallus* (BAB20424), and after each search add these sequences to your clipboard. At the end you should have five sequences on your clipboard.

Click **Clipboard** on the blue menu at the top of the page to see your sequences. It should display the summary for each of the five records. Toggle the Display window to **FASTA** and Click **Display**. Now you should see each of the files in FASTA format. When converted from its current HTML format to text format it will be a suitable multiple FASTA input file for a multiple alignment program.

We have to export this back to our own computer as a text file. At the top of the page, toggle the 'Send to' window to **Text**, and then click **Send**. Now the file consists of identification lines (starting with the > symbol) followed by the sequence in each case. There are no spaces between the entries. **Highlight, copy,** and **paste** the sequence file to a Notepad page on your computer, and then **save** it.

Open the file with Notepad or a similar text editor program on your computer. Towards the end of the identification line for each sequence is the name of the organism in brackets. We want to keep this information and eliminate all of the accession numbers, etc. For each of the identification lines edit the information behind the > symbol until only the name of the organism remains. This information will be printed out on our phylogenetic tree and shorter is better.

To run our multiple alignment, go to the ClustalW program at EMBL at **http://www.ebi.ac.uk/clustalw/#**.

This is a somewhat complicated web page, but we do not have to concern ourselves with all of it. Most of the windows can be left on their default settings. We do have to fill in several of them, however, and load our sequences.

First, create a title for your alignment in the upper left-hand window. Toggle Color Alignment to **Yes** in the top row. In the fourth row, toggle the Tree Type to **Phylogram**.

Then push the **Browse** button at the bottom and locate your PCNA text file on your disk. This will upload your file to EMBL.

Last, push **Run**.

Scroll down the output page and you will see the multiple alignment of your files. The most similar sequences are on the top and the most divergent on the bottom. Not surprisingly, the chicken and human are closely related and on top and the plant sequence is at the bottom.

The color scheme identifies residues by properties and allows you to easily see the regions that are most conserved. Below each column the (*) identifies residues that are identical in all five sequences and the other symbols give degrees of similarity.

An alignment of this type is the starting point for doing the phylogenetic tree. To build a tree, a program such as Phylip quantitatively scores the similarity between each pair of sequences based on this alignment. In this case, the tree building program has already been run.

If you **scroll** further down the page, you will see a simple diagram displaying this information in graphical form. The length of the lines between organisms and branch points are related to the degree of similarity between the sequences in the alignment. Based on our assumption of constant mutation rate over time, these distances are then proportional to evolutionary distance.

You should **save** this output page to your own disk or print it for your use.

This has been a simple exercise to look at Taxonomy resources and the basics of multiple alignments and their use in tree building. Keep in mind that there are many assumptions underlying such exercises and many alternative ways of doing this type of analysis. Ultimately we satisfy ourselves of the validity of the result when multiple methods give us a similar result. For instance, we might expect to get a similar tree if we used a different protein from the same set of organisms.

Try running an analysis on your own. Choose a protein and make a new multiple FASTA file. You might choose a different protein for the same set of species that we just analyzed. Do you get an identical tree?